FEEL THE BURN

A SOLDIER'S JOURNAL
DURING BASIC
COMBAT TRAINING IN
CHARLIE 1-13

SGT PAUL ASKEDALL

Find us on Facebook at:

www.facebook.com/alphaomegaprods

First edition 2017

ISBN 978-0-692-84149-5

The views and opinions herein do not reflect those of the United States Army and are not being used as propaganda or as a recruiting tool. The views and opinions expressed in this book are those solely of the author, and the author has not been paid by the United States Army or any other organization to express or publish these opinions.

Praise for My Hagakure

This book will never be pigeonholed into any category. It becomes clear after the first couple pages, that you are reading something that you can't deny; it's raw, it's real, it's the human brain broken down to the most simple, honest, uninhibited words on paper. You can almost hear the voice of Paul as he pours his soul into every word in this masterpiece. This is the first of many for this new genre, it is handwritten and that is only one of many things that separates this book from the others. Good luck finding a better coming-of-age journal of ideas and truth like this. I would recommend this book to anyone with a heartbeat.

- Josh C.

The great thinkers of Old wrote their masterpieces in convoluted language, using terms, phrases, and ideas which, today, we may not understand. Paul Askedall belongs in the gilded age of yesterday - but he exists today. He is a clear thinker - level-headed, not too extreme, but he provides a unique insight into the nooks and crannies of the world of today, the reality in which encapsulates us, rather than the one we encapsulate. I have read this work in its entirety; I have only good things to say about it.

-     Jeremy M.

So as soon as I started reading, I noticed that there was something different about it. It touches on a lot of things that we see and experience in our own lives but it gives a unique perspective on it. And everything else that the book explains, the stuff that not everybody experiences, really piques your curiosity. And that's a feeling I feel that we experience less and less as we get older.

-     Randy C.

Paul is an amazing young man. His memoir is both interesting and motivational. It caused me to start writing down my journeys in life. Looking forward to part 2!

-     Rachel L.

Praise for Feel the Burn

Paul is a warrior and scholar, and because of this, he's able to break down, in necessary detail, everything that comes with U.S. military training—from triumphs and instruction, to learned lessons and personal follies. It's a quick read and any new recruits getting ready for Basic would be well-advised to give it a read.

> - Michael Anthony
> Author of Civilianized: A
> Young Veteran's Memoir, and Mass
> Casualties: A Young Medic's True Story
> of Death, Deception, and Dishonor in
> Iraq.

This book is dedicated to Ryan "JR" Gersch, whose love of life and others has inspired me to be better. Keep the light burning, brother.

This is also for Ben, who was a warrior in every sense of the word.

FEEL THE BURN

A Soldier's journal during Basic Combat Training in Charlie 1-13

SGT Paul Askedall

Feel the Burn

"I will sweat more in training so I bleed less in combat."

-Dick Marcinko

Co-founder of Navy SEAL Team Six, taken from the Ten

Commandments of SpecWar

Introduction

The military is not for everyone, plain and simple. If someone

thinks they'll get an enlistment bonus and Active duty pay easy, think

again. They see the numbers and figures of pay based on rank and they

look into the different jobs, find one, and then assume they'll do great,

along with how much they'll like or dislike their MOS (military

occupational specialty). You have to get through Basic Combat Training

first. You have to survive BCT (nothing compared to other Army schools

like Ranger school, which is probably like BCT on steroids, an energy

drink intravenously draining into your veins every hour on the hour,

while fighting ninjas as you simultaneously fight sleep and hunger....but

1

there are people who cannot even get through BCT), and when you get into the winning and succeeding mindset and create a goal and the attitude and iron will to make that goal a reality, your body follows through. Set no limits, aim high, and you'll be surprised with how much you can accomplish and how far you reach. Personally, I liked basic training. I enjoyed the hooah, balls-to-the-wall field training and learning crisp execution of drill and ceremonies. Some of the platoon punishment was hard, but in hindsight, it was great and I am glad I experienced it all. I am a warrior, being a part of martial arts for many years before joining the Army, living by a martial philosophy and applying a strict set of moral principles.

Discipline is nothing new to me since I've lived with it from when I was real young; between living the martial Way and being physically abused by a strict stepdad, I am accustomed to having to do something that is painful, pushing my mind and body further. My stepdad was a strict Englishman raised in Australia, and he brought that strictness to the household when he married my mother. As a child, I had to kneel on dry beans on concrete in shorts while I put my hands on my head, and if my arms dropped slightly from being tired, I was slapped with a yard stick and told to kneel longer while holding my arms up for an increased length of time. Another punishment I had was being forced to write the words and definitions from a beginner's dictionary, A-Z in 20 minutes. If I didn't finish the whole dictionary in 20 minutes, I was hit and told to restart the task in a shorter time. This was humanly impossible, especially as an 11 year old. You can tell I was hit excessively. Recalling another time that wasn't so bad, I had to stay out in the backyard the whole day until dinner, but I didn't mind that too much because I had an imagination, a swing set, toys, treehouse, tools, and dirt. During this time in my past, I wanted to absolutely kill my stepfather, but now I am significantly older, acquired a philosophy on life at an early age, and I no longer hold a grudge. I never knew it at the time, but his punishment and verbal abuse helped to form me into the person I am today.

My sometimes shitty-at-the-moment childhood living with him helped carve the path to me taking up martial arts. I got into the arts for self-defense at first but over the years it became a way of life, bettering myself and helping to better the people around me. The code of Bushido, what many samurai of feudal Japan lived by, has seven virtues: Honor,

Respect, Loyalty, Selfless Service, Duty, Courage, and Sincerity. I live

by these values as best I can and I encourage them in others. One thing I

found funny was that these seven values of Bushido are the same seven

Army values taught to soldiers. Yes, the U.S. Army holds the same

virtues as the warriors from a nation we fought against in the past, but

nobody monopolizes on noble ethics. Any individual who truly adheres

to these values or a strong ethical grounding sets the foundation for

substance in the present moment, an impressionable future, and paves the

road toward a better world. Returning to the matter of discipline, for

those individuals that severely lack in that category, and if they do not

have open minds or the ability to adapt to new situations, they will at

least get a rigorous workout and clarifying experience before they

graduate basic.

I am a great conversationalist and in my platoon there were males and

females of different ages from many different states (even different

countries), which made for good discussion. Being from California and

living in New York for nine years, you would think I wouldn't care too much for other states, but I've traveled to a few and I'd like to know about others; the certain lingo and habits of the people of a certain city or town, and what it's like to live there. You can also guess that living in both states on both coasts has made me somewhat jaded. It's almost like a different world in another state, if you compare Southern California with, say, Alabama. There are a large number of recruits who get to basic training with an ego, selfish and arrogant in their way of living around the barracks or in talking and dealing with others. That type of personal pride does not get someone far in the military, especially not in basic. We are trained to work as a team; there is no individuality in the military, and Drill Sergeants stress that enough. If one or two soldiers in your platoon fuck up, the rest of the platoon pays for it. So it should be in your best interest (and the interest of your group) to not fuck up, or to help others stay on point and not drop the ball. As much as I like to stand out from the crowd, as much as I like being unique, I was assimilating into a sea of green digital camouflage ACUs (Army Combat Uniform) with a pair of my own.

Section I: Hurry Up and Wait — Reception Battalion

Finally. After waiting three months in the delayed entry program, eagerly wanting to leave and start Basic Combat Training, I am at an airport waiting with another recruit, having lunch. We're spending the thirty minutes we have until our connecting flight to South Carolina eating and talking. Man, I love airports. The atmosphere, the people from many different states or countries; it's a place I would spend all day in (and unbeknownst to me at the time, I really would spend all day or all night in an airport, both because of the Army and my personal travel. Upon arrival to the South Carolina airport, we were directed to the USO, and might I say, the USO in the South Carolina airport is nice! Computers, big screen tv, couches, drinks, and snacks---everything you *won't* have during training. Better enjoy it...but I didn't, one, because I already ate, and two, I didn't want to put anything into my body if I had to do PT (physical training) later.

As soon as the charter bus stopped inside Fort Jackson, all of us on the bus waited eagerly for a Drill Sergeant. We were tense, anticipating the coming storm and as soon as our first real-life, in-person Drill Sergeant walked towards us, we all shut up right away. Falling into our ordered formation and being briefed on the first few hours of orientation, I remember one thing specifically; the Drill Sergeant pointed to the crest on the right ACU jacket breast pocket and said, "The best way to distinguish a Drill Sergeant from a regular noncommissioned officer is this. This is Drill..." then he points to the Staff Sergeant rank insignia on his chest, "and this is Sergeant."

Points to crest. "Drill..."

Points to rank. "Sergeant. Drill...Sergeant. Drill Sergeant."

We had no sleep for the first three hours while we were issued PT uniforms, threw out any contraband we had in our civilian bags, and were given hand sanitizers and what's called a "smart" book, which contained Army history, explanations of the different legal actions that can be attached to one's military record, and textbook training/descriptions on everything from firearms to squad formations, proper attire to subordinate/superior etiquette. Going to sleep at 0300 and waking up at 0500 was a killer, and it was more of the same shit from the night before: more briefing and more issuing of items. This time we got

our free haircuts, courtesy of good ol' Uncle Sam. After living with

spikes for years, it felt alien to run my hand over my head, the fresh buzz

cut signifying a new life...and in a very large way, it was. When we were

in line being issued our ACU's, boots, and patrol caps, we were stuffing

everything into our enormous duffel bags. Just when I thought there was

no possible way anything else could fit into this stuffed bag, I was

proven wrong. You *can* fit a whole lot of shit into that Army bag. Plus,

when you think there is absolutely NO WAY something can be done, the

Army pushes through that doubt and makes it happen. The first time we

got our ACUs everyone was psyched, and some were even working out

in them. In Reception since we still had our civilian backpacks, I took

out my camera and snapped some pictures and shot some video, not only

for memories but for documenting my experience for my "Path to

Becoming a Soldier" video. I've met a few recruits from both California

and New York, and of course I had to ask them what town or city they

hailed from.

Standing in line or sitting, waiting to receive our next item, waiting to be called, I've learned that in the military personal space, your own personal "bubble," doesn't exist.

"Squeeze in, make it tight."

"Get in close, nobody's going to bite you."

"Close the gaps, make room for people after you."

It's almost like an assembly line, except the machines in line get yelled at for talking or not having their uniforms looking sharp, i.e., shirt not tucked in, collar not down, a name tape in the wrong place. Tis common sense to dress proper, ESPECIALLY in the military. Not only do Drill Sergeants stress professionalism and decorum, you are becoming a soldier of the United States Army, and the U.S. military is all about uniformity and standards. You are no longer on the streets, no longer within your safety zone, comfortable and complacent. You are no longer living on your time. In uniform, you represent something greater than yourself. You represent a whole nation, and hundreds of years of military tradition.

Getting shots wasn't bad, although I did feel every single one of them and I did taste the nasal spray for the swine flu vaccine. My shoulders were sore for a good four or five hours.

My Reception platoon was 81. We were a good group and shared the bay with part of 80th platoon, many of which were pretty cool. There was a tall guy from Washington who was a gunsmith who also joined the Army as one. We talked about firearms but he definitely proved his knowledge over mine. Oh yeah, totally schooled.

Prior to BCT many people had hopes of not getting chewed out by a Drill Sergeant and they believed they would do the best they could as to not attract negative attention. I don't think this one guy had thought about that at all. We, the platoons, called him Turtle because the very first night we arrived, the VERY first night, a Drill Sergeant told him to walk fast-- repeatedly. After receiving our PT uniforms we were told to walk at a brisk pace down the hallway and fill in the space after the ones that sat down already, and Turtle made an effort to walk fast but it wasn't fast enough. So the Drill Sergeant made him walk back to where he was standing to try again. And again. And again. And yet again. Receive, Walk, Repeat. Turtle just did not get it through to his head. It was pretty funny.

Day 2 of Turtle frenzy; 81st platoon was getting financial documents taken care of and pictures taken for military IDs, and a sergeant told recruits who were waiting to sit down, and when we all sat at the same time, Turtle was the last to sit. Here we go again...

The sergeant made us get up and as he saw Turtle was the last man again, there were a good handful of times he made us sit down, stand up, sit down, up, down, up, down. Turtle was slow and he was definitely noticed, this time by another NCO. He was asked by the sergeant what his problem was, why he couldn't follow a simple command and sit and stand in unison like everyone else. The sergeant also asked Turtle where he was from. After stumbling on his words and half mumbling, half stuttering the answers, Turtle stated he's from Wichita, Kansas. The NCO laughed, turned to his NCO buddy and said, "You hear that? He said he's from "Wichi-taw, Kan-sas." He emphasized the city and state with a heavy southern drawl imitation. Throughout this whole charade, I tried hard not to laugh or even crack a smile---I was right next to Turtle. When we were getting financial documents in order for BCT, I showed a representative my driver's license and she commented on my spikes--- the spikes I used to have, at least. I told her I've been called a Japanese anime character, and that I enjoyed taking the gel and shaping it how I wanted.

Any time we finished eating, 81st had to fall in formation behind a World War II tank by the DFAC (Dining Facility). While waiting for our next move everyone would watch formations march by, singing cadences. At the time, I didn't care about cadences, I was impartial to whether we sang them or not. I just wanted to get to the fun stuff; drill and ceremonies, firing weapons, combatives, obstacle courses...just a few events among the many listed in the Army brochure.

The recruits of 120th Reception Battalion were finally on the way to Basic Combat Training. We turned in our linen, packed everything we received into our monstrous duffle bags, including our civilian bags--- don't know how but we did---and lined up in company formation for accountability. Upon checking off everyone and making sure nobody ran off, we started loading our laundry bags (also full) onto a military truck called a 1-ton. A few platoons marched to the buses and once everyone filed onto my bus, one Drill Sergeant stepped on and immediately started yelling.

"SHUT UP! What you're going to do is stick your fuckin' head in your bag, eyes staring straight. I don't want to see ANY eyes, nobody fuckin' look up. And I don't want to hear any fucking talking!"

This was it. This was the second moment we've all been waiting for, the ride to our BCT company. I'm not going to lie, I wasn't nervous. Mostly eager again, anticipating the screaming, the rushing, the punishments, and tedious schedule. Between the silence inside the bus and the muffled sound of the engine outside, I could hear my heart beating. It wasn't beating fast but it was one of the only sounds in my head on the way to a place I thought I would almost never be.

Section II: Push Forth with Effort — Red Phase of BCT

"There is one of two ways you will get through basic Pri'ate; you will either get smarter or stronger."

-    Drill Sergeant Price

I want you to think of the scene from Saving Private Ryan that portrays D-Day, when the Allies stormed Omaha Beach at Normandy on June 6, 1944. The quiet boat ride all the way to the beach, the unnerving anticipation of impending death and destruction, thoughts racing through the minds of the soldiers behind their silent, contemplative faces. A quiet ride to the beach and then, as soon as the bay doors drop, the air explodes into a hail of gunfire and yelling. That is the analogy of how it was on that day of arriving at Charlie Company, 1/13th Infantry Regiment, minus the gunfire and explosions. The trip was quiet until the door opened and the Drill Sergeant told us we had thirty seconds to get off "his fucking bus."

Everyone jumped up and hastily headed off, and "Day Zero," as a Drill Sergeant called it, went off like a bomb. Drill Sergeants screaming at

14

everyone, barking where to go, boots pounding the sand and grass like at a horse race. I was the last one off my bus and I was really hoping I wasn't going to get chewed out for being so, but as soon as my foot hit the ground I bolted with my duffel bag to the company formation area and my bunk mate from Reception tripped, his bag falling out of his arms and him heading to the ground. A Drill Sergeant asked him, "Why are you stopping!?" And I was stifling a laugh and a smile as hard as I could because of two reasons. One, he asked why my buddy was stopping, when he obviously tripped involuntarily, and two, because I wasn't chewed out for being the last one off the bus and ran right by the Drill Sergeant. Success!

We lined up in formation while the constant screaming was going on, then we were told to retrieve our bags from the pile that was accumulating in the middle of the company area (a select few recruits were ordered to unload the truck). We had five minutes to find our bags and when all of us knew we weren't going to find them in time, we started shouting the names on the sides of the duffels, helping the group that much more. Finding our bags, everyone returned to their designated formation areas where we then emptied the duffel bags and took inventory of our clothes and items. After, we ran up to our designated bays and were issued a bunch of equipment; ACH (Army Combat Helmet), IBA (Interceptor Body Armor), LBV (Load-Bearing Vest), M16 magazines, assault pack, 2-quart canteen, and a large rucksack.

The first week of Red Phase was tiring. Wake up was 0400 followed by an hour of PT, followed by constant sessions of getting smoked...followed by two minute meals...followed by more smoking. That was followed by classroom briefings, which is difficult to stay awake in because of the small amount of sleep, or "rest" as Drill Sergeants call it, coupled with the continuous flow of always doing something. Add a warm classroom to that and you have a recipe for disaster.

Little sleep that my body was not yet used to: check.

Almost nonstop activity from wake up to lights out: check.

Getting smoked for falling asleep during class: fuck that.

There were a few times I would nod off, even warned by another recruit to wake up, but never have I full-on fell asleep in class, which would result in being smoked. One time I nodded off and a very attractive female recruit from another platoon sitting in a desk next to me nudged me awake with her shoulder. My eyes opened, I turned to her and thanked her with a whisper. She nodded.

One morning wake up was at 0430 and all of us in 3rd bay decided to wake up and be out of bed getting ready at 0400. Coincidentally, one of our platoon Drill Sergeants happened to want to barge in and get us up--- at the same time. So now, to her, it looked as if we were sleeping in.

On the PT field before stretching, the company would recite the Soldier's Creed which I already memorized before basic, along with the general orders. The stretches were not as long or substantial as the martial arts stretches I've done, but they did help with the conditioning drills and exercises we conducted.

The Front Leaning Rest

There are two commands given; the preparatory command that lets you know something's coming, then the command of execution that is carried out. Standing in formation, one specific preparatory command and command of execution many recruits dreaded was, "Half-right, FACE," because directly after we all knew what came next: "Front leaning rest position, MOVE." The front leaning rest position is a push-up position, holding yourself up while waiting for the next command, which was usually to count in cadence with the Drill Sergeant, and the next step up from a regular front leaning rest is elevating your feet on your bunks or something similar. Recruits are told to elevate their feet either as an individual punishment or when the whole bay is getting smoked. Doing a push-up is also called "beating your face."

The company was finally issued our rifles. They are called rifles or weapons, not guns, but when I was taking my hunting license qualification course when I was younger, the class was told never to call a gun a "weapon." The instructor stated a firearm is only a weapon when pointed at another human being. Well, welcome to the Army, where a soldier's duty is to serve his or her country, and that means potentially pointing a loaded firearm at another human being.

While looking over my rifle, I noticed it had scratches all over it, and as I looked at my platoon's rifles I saw most of theirs almost new, with no superficial blemishes and a smooth matted finish. Granted a large number of these weapons are many years old compared to the newly issued ones, but the condition also depends on how well the previous owner maintains it. I couldn't wait to shoot it; the last time I fired a rifle was at Boy Scout camp, many eons ago. A .22 Marlin was what I fired but that's nothing compared to an M16. I really wanted an M4 instead of the musket but the M4s were only given to smaller/shorter people. This was the *only* time I envied a small person---other than being small enough to crawl through a McDonald's Playplace. But it wasn't bad, I really didn't care. I had a rifle and this one was mine.

On the first weekend of being in Charlie 1-13 my body was adapting to the shots we were given and I got sick along with a handful of others. We were standing in formation after breakfast and I really wanted to lie down, my head in a really bad place. I then felt my throat salivating, readying for what was to come up. I ran out of formation, found a grassy spot around the corner behind my platoon, and hurled. A moment later a concerned buddy walked up with a Drill Sergeant.

"Why are you vomiting? You need your nutrition," the Drill Sergeant asked.

I spit and replied, "My apologies, Drill Sergeant, just a minor setback.

Won't happen again, Drill Sergeant."

"Do you need to go to sick call, Private?"

"Negative Drill Sergeant."

"Are you sure?"

"Yes Drill Sergeant," I reassured him as I spit some residual vomitus

from my mouth.

"Alright, get back in formation when you're up to it."

"Roger that, Drill Sergeant."

I spit again into the grass, slowly stood up, and returned to formation.

For a few more hours I did the best I could with my tasks until the

headache and nausea subsided.

In the beginning when the bay was cleaning every day and we were

sweeping, mopping, or cleaning the tops of lockers, I was always the one

to pull my own weight and then some. When some other recruits swept,

they wouldn't get under the bunks or between the lockers and that's

where most of the dust bunnies congregated, thus leading to the floor

always having to be swept. We then realized that no matter how much we swept at any time of the day, the floor would always be dirty. Back to pulling my own weight, I would always pick up cough drop wrappers and other small pieces of garbage while on the way to the latrine or around someone else's bunk. My assumption was, if others saw trash they would think, "I don't have to worry, fire guard will sweep it up in the middle of the night," or they saw it and, plain and simply put, they have no discipline to pick up a tiny wrapper or tissue and walk to the latrine and throw it away. They simply couldn't be bothered. Lazy bastards. It gets really fucking tiring having to pick up after others…constantly, especially at an age when these adults are supposed to be men. I emphasize the word constantly because people are consistently leaving shit lying around. I have much patience but after weeks of cleaning and rarely getting any time to myself, small shit like that adds up and annoys me, like an itch I can't scratch. It's irritating. Dust bunnies, heh heh. That's cute.

Returning to cough drops, they're a currency here in basic. One battle buddy of mine, Stewie, acquired $140 just by selling them from every package he received. That's insane. If he was running low on the regular menthol-flavored cough drops he would sell one for two bucks, and then a Halls Breezers (the fruit-flavored drops) for five bucks if he was low on those. Damn, it was like cigarettes being a currency in jail. People also traded them for other items or passed them out when they had an abundance. Personally, I have never had so many cough drops in one day. I never eat cough drops back home, and I hardly eat them here for a cough; they are good mints after a meal and they're like candy. Drill Sergeants said not to eat them like candy but everyone did. Cough drops were the only sweets we had aside from the candy in the MREs, because not many recruits wanted to be caught eating pie or other desserts in the dfac. I didn't care; I would just burn it off anyway. I was losing weight with the whole constantly active part of basic in addition to morning pt, so it mattered to me not if I partook in sugary delights.

As the days dragged on, there was more pt, more smoking, more of my body adjusting to the schedule. The company was given "smart sheets" and "smart cards," two sources of study that are very important to basic combat training. We had to learn everything on them and much more info and facts in the books we got from 120th. As if we didn't have enough stuff to worry about. When are we ever going to get the time to learn all this?

2000 hours: mail call. At this time you have the freedom to finally do your own shit like write letters, personal pt, talk, take showers, what have you. I spent much of my time writing this here journal, other times I would write letters and do more pt. As a result of getting smoked for not having the bay clean, we even cleaned the bay and latrine on our personal time. Some of us did, anyway. When mail came around it was 10 push-ups for a letter and 35 – 50 for a package. We all looked forward to personal time, and everyone looked forward to getting mail—everyone except myself. I had a handful of addresses to write to but I didn't start writing letters until later. I was too busy with everything else and I didn't quite miss everyone from the civilian world like everyone else did. I wanted to get away from it for a while and I was cut off from the outside anyway, by all but a letter. I had been keeping everyone I knew in mind though. The company was only allowed two phone calls ; one in the beginning to let our families know we made it and we weren't dead, and the second was to confirm if anyone was attending our graduation. We never touched a computer for the duration of training, and that was nice.

It was a good break from social media and staring at a screen for hours on end. To this day when I hear any other service member talking about their basic training experience and how they got to use their cell phones and use a computer, I scoff and say they didn't have hard training.

I came to realize Red Phase is about pain; it's about sweat, pain, forming discipline, breaking people down, and more pain. Our company had a longer ruck march to a location for a 3-day FTX (Field Training Exercise). I volunteered as "road guard" again---four recruits don reflective vests and they block traffic so the formation can march safely across the streets and intersections. The road guards are the "blinkers" in a formation, referencing the front and rear lights in a vehicle that signal the driver is turning left or right. The only difference to being a road guard this time was, we had rucksacks. The road guards in the front would run to the crossing to get there before the formation, and the ones in back would take over then run to catch up with the formation. I didn't mind being a road guard; hardly anybody would step up to do it because it required two things. One, to put yourself out there and show commitment and teamwork to the Drill Sergeants, when most of the company wanted to stay in the background, not be noticed, and "just get by." Two, being a road guard involved being more physically active than the rest of the formation on marches to and from a location because you are constantly running---again, no problem for me. It really only sucked on warm, sunny days because I warmed up real quick, especially with a rucksack and weapon. When we arrived to our campsite, we downloaded our gear and quickly set up our one-man tents. Back at the company area before we set out for this FTX, one of our Drill Sergeants was getting pissed and told the platoon we had fifteen minutes to set up our tents or we'd get smoked. It didn't take fifteen minutes to set up one of those tents, but some individuals just couldn't get it set up. Nothing like a little incentive to speed things up! That night word got around that the Drill Sergeants were going to see how many rifles they could acquire during the night, so we all slept with our M16s and M4s in our sleeping bag. Not much space to move around and not too comfortable, but you make do as best you can with what you have. Besides, enduring a little discomfort was well worth it compared to the consequences one would face if one lost his/her rifle.

We awoke at 0400 hours and it was cold. I didn't have hot water in the field so I learned to save the MRE heaters to heat the water so I can use it to shave. I lacked a mirror too, but I didn't let that stop me. I knew my face. For breakfast we had semi-warm food (I say semi-warm because the scrambled eggs were warm but the bacon and sausages were cold) and orange juice or water. Good ol' Army chow! Over the next three days, we learned different tactical formations, grew real familiar with house clearing, and were trained in vehicle searches, as well as practiced more individual and buddy movement techniques. All of this was great training but we were all exhausted, and the Drill Sergeants kept up the motivation and momentum. How did they do it? Some of the recruits were thinking they were robots or on crack. I eventually grew to learn as a soldier, you make friends with coffee or energy drinks, or dip (chewing tobacco), which I never touched. I even heard stories about recruits taking the MRE instant coffee packet and emptying the contents into their eyes or clumping it under an eyelid. I began to wonder how much of these stories were eyewitness accounts or rumors.

On the morning when the company was expected to head back, the Drill Sergeants specifically demanded none of us wear our polypro (polypropylene cold-weather base layers worn under uniforms) tops or bottoms on the ruck march back. They said we *will* warm up with simple ACUs and ruck, so no matter how cold we were at the moment, they told us not to wear polypro. Drill Sergeant Price joked with us, saying we *could* wear them, but we would suffer for it, not only by our own stupidity, but the wrath of the Drill Sergeants too.

Making the Walls Sweat

Charlie Company started rucking it. Halfway home, a couple of recruits started getting hot, and one almost passed out. One of those recruits--- and the same one who almost passed out---was none other than my buddy "Chrizzle." His face was red and he was sweating profusely. Drill Sergeant Price was moving down the rows of recruits during a break, yelling at us, proving to us what he said was not a suggestion. Chrizzle just became a blue falcon (someone who screws over his or her buddies for personal comfort or gain).

27

When we made it back to the company area, each platoon was told to go to their platoon bays after unloading our gear. After 3rd Platoon was finished and filed into 3rd Platoon bay, Drill Sergeant Price demanded we toe the line. The platoon lined up in the bay, what is called "toe the line," where everyone touches their toes to the same line on both sides going down the whole middle of the bay. The middle of the bay was called the "kill zone," where recruits were not allowed to cross into, which was relatable to keeping off the grass in many areas on base. We lined up and knew we were fucked. Drill Sergeant Price slammed the bay door shut, and then practically threw the white board across the front of the bay.

"What the fuck did I tell you, Pri'ates?! I said don't wear your fuckin' polypro on the ruck march, Pri'ates. Why? Because ya'll are gonna warm up on the march. I fuckin' told you ya'll would warm up, didn't I?"

"YES DRILL SERGEANT!"

"Well because one of your battle buddies decided not to fuckin' LISTEN, ya'll are going to suffer for it, Pri'ates!"

Drill Sergeant Price walked up and down the bay, in between both rows of recruits, furious.

"When us Drill Sergeants tell you to do something, and you don't listen to what we tell you, and it almost gets you killed Pri'ates, our boss, the First Sergeant, gets fuckin' pissed at US. So I got chewed out by the First Sergeant, and you know what, Pri'ates?! Shit rolls downhill. I'm handing it back to you Pri'ates! Ya'll are gonna feel it. Front leaning rest position; MOVE!"

The platoon got down as commanded.

"In cadence."

"IN CADENCE," we echoed.

"Exercise!" Drill Sergeant Price counted cadence extremely fast, forcing us to keep up.

After over a hundred push-ups, we elevated our feet on our bunks and did more push-ups. Now, doing any exercise in cadence with a count makes you perform two of the exercise. We got really high in number and were feeling it…and we were far from over.

"Turn over, Pri'ates."

We did as we were told.

"The Flutterkick!" boomed Drill Sergeant Price's voice.

"THE FLUTTERKICK, DRILL SERGEANT!"

As we lay on our backs, we scissored our legs vertically as we attempted to keep up with the hasty count. On my own, I can do a hundred flutterkicks without my boots...but these were in cadence and high in number, *with* boots. The platoon's arms and shoulders were hurting and fatigued, and now our legs and hip flexors were tiring and giving out, many recruits wanting to give up.

Drill Sergeant Price walked up and down the bay, still fuming but this time his anger was refined, calm but stern.

"I'm going to smoke the dog piss outta you, Pri'ates. You're going to sweat so much, it'll feel like you're pissin' out your pores." *Imagine that, I thought, sweating so profusely it feels like you're pissing through your pores.*

Drill Sergeant Price continued, "We're gonna make the fuckin' WALLS sweat, Pri'ates. All the windows, the doors---all shut. You're gonna feel it, Pri'ates."

And feel it we did. Two long hours of getting smoked, bad. We did the supine bicycle, the push-up, the sit-up, squats, military bench press---all with our rifles---v-ups, flutterkicks, 8-count push-ups (another name for burpees), and we did a duck walk around the bay, "nut to butt." Our whole bodies were burning and sore, and we were exhausted physically and mentally. Many of us wanted to throw a blanket party for Chrizzle. You remember that fat kid from Full Metal Jacket who always fell short, which cost his whole platoon a smoking or extra duty every time? And the scene that was an inevitable beating with a bar of soap in a sock that left no bruises on him? That scene many of us wanted to exact upon my buddy.

I'm sure this is another rumor because my company sure as hell didn't have them, but me and a handful of other recruits started hearing about stress cards. A red card that a recruit can take out and hold up in the air signaling a "time-out" from a smoking session or training due to the amount of stress one is enduring or the lack of fortitude one contains. We never had these so-called stress cards, nor did we ever see them in other companies. I'm sure they were something made up as a story to tell about other recruits, emphasizing the fact that the military is not for everyone. I was annoyed by the sheer idea of them. There is no war in this world that would stop momentarily for any soldier because they "couldn't take it" or they "needed a break." Fuck that bullshit, no organization should have a say in how wars are fought, outside of the Geneva Conventions and

Law of War, which state certain weapons that cannot be used in wartime and underline the treatment of prisoners of war. A damn stress card should never be anywhere near a written policy or approved for application. No battle will have combatants that abide by your request to cease firing based solely on one's mental and emotional health. Great military leaders of history throughout the world would laugh at such an idea.

*   *   *

Before grouping and zeroing our rifles on the actual range prior to qualifying, the soldiers had to practice getting used to the fundamentals of Basic Rifle Marksmanship, so we marched over to an indoor range that mimicked a real range. This place had enormous projectors that showed a 15-lane range against the wall. These specifically-designed rifles and electronic range were the closest I would get to video games in three months, and it was fun. Too bad there were no ducks flying across the wall or a hound dog running back and forth! Okay, for those of you who don't get that reference, it's a Nintendo game, from *way* back in the day.

Waking day after day, early and habitual, there would be days I just didn't want to get up. Having slight pain, being exhausted and growing tired of the day-in, day-out routine, I never wanted to quit but at times I questioned why I was there. I wanted to sleep in, not only to get more sleep, but because I was so damn tired of waking up so early. What I would give for more rest…but whenever I *thought* of quitting, I would remember the reasons why I joined and the honor of staying true to those very reasons. I would also bring to light the fact that I was getting paid to do this. Besides, I was never one to quit, I never have with anything and I never will. A line in the Soldier's Creed says "I will never quit," and I don't plan to.

Victory Tower

Hooah, I get to rappel down a 30-foot wall! Once I climbed to the top I looked down and felt excited. I leaned a little further. A Drill Sergeant from 1st platoon pushed me back.

"What are you doing, Private?! Trying to kill yourself?" *Nowhere near, Drill Sergeant, I'm amped up!* I said in my head, for fear of a smoking if I said it under my breath. I hooked the rope to the Swiss seat I had on (tightly too, it was hurting on the practice wall but for some reason not on the actual rappelling wall), leaned back off the wall, then jumped and descended. I pulled the slack to stop, kicked off the wall again, and descended quickly. I had such a great time I wanted to do it again. At the top there were people who are acrophobic but you have to overcome your fears, especially for one who is going Airborne. Here at Victory Tower you're only jumping off a 30-foot wall and secured by a rope. At Jump school on the last week you're jumping out of a plane with nothing to secure you and a parachute to break your fall. Better get used to heights soon.

Being scared of heights is one thing, but one really does not have a choice to pass on the wall, because the Drill Sergeants force you down, you have to do it, or else you can't graduate basic. So in a big way it's motivation for you to rappel down that wall. Then again, the secret of basic training is staying motivated. It's a big mindfuck. Drill Sergeants stress to the point of staying motivated, always sounding off "loud and thunderous" during formations and pt, and getting to the event of graduating. Drill Sergeants drill it into your head (oh, so *that's* why they're called Drill Sergeants! I never knew!) that *this* is a requirement to graduate and *that* is a requirement to graduate, and there is a consistent, almost never-ending forward mindset. For the recruits that lack discipline and aren't all about the physical aspects of being soldiers, it's even difficult to have fake motivation because all they want to do is leave and go back to their old lifestyle, relax and not be yelled at or told what to do. Even with a countdown of the time they have left, they no longer have their goals in view, what they signed up for should not be this difficult. Well tough shit, you signed a contract for a reason now you have to *earn* it. No reward comes without effort. There's no "free ride" in the United States Army; you want a job and money as a soldier, you have to work for it. You want to wear the uniform and feel proud serving your country, you have to suck it up and complete training. Unless you enjoy the field training and experience you get at basic, then there's no complaints! Besides, who doesn't like a challenge? If it doesn't challenge you, it doesn't change you!

Drill Sergeants say if you join the military for all the wrong reasons, you won't get far. If you join for selfish reasons you will potentially get someone killed if you are ever deployed. I met quite a few like that in my company. I never want to go to war with someone that would save only themselves, someone who is undisciplined, leaves others to clean up after them, and who thinks nobody can teach them anything new because they believe they know everything they need to know in life, which is highly doubtful. Then again, in the moment of truth, people like that can prove one wrong…but until I witness it, they're a no-go on my list.

The Team Development course teaches recruits---I'll give you one guess, it's in the title of the course---team development, which develops teamwork, camaraderie, and various approaches to problem-solving. The objective-based sections of the course are timed, and can be completed in the allotted time, as long as the members of each team come up with a plan and apply it, working together toward the given goal. My team successfully completed all but two sections. One section we would have finished if we had a little more time, and the second section, we had a few ideas but they didn't pan out as well as the team wanted. During that failed second section was one of the times I stepped up to help put the plan in motion, and I received a badly bruised (and possibly fractured) thumb.

Our objective was to make a bridge with three wooden columns and a 2-foot rope, then transport a crate filled with mission essential supplies through the first culvert, over the bridge, and through the culvert at the end of the section. We had to do this without dropping the crate, letting a teammate fall onto the ground (the culverts and bridge were only a foot off the ground), and without touching the bridge to the ground. Any of the above would result in mission failure and your team would have to start over. After projecting a possible route, my buddy Z and I decided to put it into action. He was going to lower the first column to start the bridge and I was going to hold him so he didn't fall. We started with the tallest (and heaviest) column, and as he eased it down with the rope, the rope slid down the column, Z caught it and angled the column up so it wouldn't fall, but because he did so, the column slipped from his grasp and with all its weight landed on my left thumb. How was I situated? Lying on my stomach, arms wrapped around Z's ankles to keep him stable and prevent him from falling. The impact hurt but I was grateful the hard wood column didn't land on my head. During one of the other sections I was shimmying from one platform to another and was cut as the pulley pinched my skin between itself and the industrial cable I was shimmying on. Bleeding, I thought to myself, *I guess I've taken one for the team…twice!*

I later discovered me and my team were on the basic training dvd during one of our successful course negotiations.

Fit-to-Win course. This was a fun little number; you and your team are to negotiate the obstacles of this course, and the first platoon to have all teams back first, wins. Balancing across beams, climbing up and down ropes, and crawling up, over, and through trenches and tubes were just practice as I previously came from a hobby of free running and environmental ninjitsu training.

Sweaty, dirty, and amped up from the excitement, me and my platoon waited for the last members of each platoon to finish the course, cheering and yelling as we watched every last two-soldier team come around the bend and link up with their platoon. My platoon, the Wolf Pack, saw our last team running from the last obstacle and we cheered heavily. As soon as we were a whole pack again, we started chanting loudly, "WOLF PACK! WOLF PACK! WOLF PACK!" It was awesome. We won and it felt great to share that as a platoon.

The same day as the Fit-to-Win course, we were going to immerse ourselves with Orthochlorobenzylidemalononitrile, more commonly known as CS gas. Say *that* five times fast! This tear gas is stronger than CN gas and is widely used during riots and demonstrations. The thing with CS gas, it reacts to moisture on the skin and eyes. And we just finished the FTW course. What fun.

We lined up outside the gas chamber in rows and went over gas mask clearing procedures. Once my row went in, the door closed behind us and a Drill Sergeant was in the middle of the chamber, waiting. We were told to lift our masks, breathe in, and then clear the masks. One kid from my platoon, who was a follower and very egotistical, caught a load more gas than he wanted and he wasn't properly clearing his mask so he panicked and tried to run out. His head shifted quickly to both doors, seeing which one he should escape out of, and when he attempted to make a run for it, one Drill Sergeant grabbed him and shoved him back against the wall. The kid went limp, I'm guessing it was because he fainted. I laughed; that was amusing, him acting so tough then panicking when faced with an assumed danger to his life. BCT is a very controlled environment, with Drill Sergeants monitoring every situation and soldier as much as they can in given exercises. This kid was not going to die.

We took off our masks for the last bit of the chamber and sucking in a shitload of gas, we tried to recite the Soldier's Creed. I only got to the third line before I started coughing uncontrollably. The sensation was as follows: stinging in my lungs, nose, and throat, eyes watering, and I was burping but tried to take in more oxygen but couldn't. I'm in this point of limbo, trying to breathe but burping at the same time, then coughing when I think I have more air.

Before basic, I read a book about preparing for BCT with what to expect, what to do and what not to do. One of the things on the "Do not do" list was touch your eyes during the gas chamber, so I remembered to heed that bit well. But it was true what's said about the effects disappearing almost instantly once you walk out into the fresh, uncontaminated air. As some soldiers threw up or fell as they stumbled out of the chamber hastily, others had snot dripping from their nostrils and hanging like a piece of melted cheese. I was among the group of others who were lucky to have runny noses and tears, but no snot. I was glad I blew my nose earlier that morning! I didn't get that from the book, it was just a common sense precaution I took.

When we learned more Drill & Ceremony I acquired a great sense of pride; proud to be part of a decades-long tradition and proud to be contributing to the discipline and professionalism of being a soldier in the U.S. Army. I enjoy taking part in D&C, and I aim to always make crisp, precise movements as I execute the command given, and *dammit* do I love saluting. It's a great honor and respect to salute an officer, and living by knightly values, I like to carry on the tradition of the salute.

One theory of the origin of the salute is, knights lifted their visors with their sword hand, showing each other respect and honor, as to say, "This is who I am, you see me as I see you." In relevance to precise execution of movement, there is a type of Buddhist meditation called oryoki, which means "just enough." Oryoki is a meditative form of eating and emphasizes mindful awareness by performing a strict order of exact movements.

Section III: By Endurance We Conquer — White Phase of BCT

Many smoking sessions and Drill Sergeant pep talks later, my platoon and the company finally make it to White Phase. To tell the truth, White Phase felt no different from Red Phase. We still got smoked badly for being too slow or paying for a platoon member's mistake (or stupidity) and we still took frequent trips to the range to get better acquainted with our rifles.

Every platoon had a female Drill Sergeant in addition to two males, and I was told by recruiters, Army internet forum members, and soldiers from my Reserve unit that female Drill Sergeants are the worst. Just my luck---I wanted hard training, and I was going to get it. Trust me when I say this: she was scary. A small, black female who knew how to raise her voice and when to use it, everyone tried not to get on her bad side. Some may say female Drill Sergeants are the worst because they have a great deal to prove. Whatever the reason for their attitude was, we didn't care. We just tried to stay out of her line of sight or out of range of her voice, which was impossible for either. We actually felt sorry for the other platoons when she went off on them…almost. Me and a bunch of other buddies laughed every time she started yelling and smoking another platoon. In the field other platoon members came up to us in 3rd and sympathized, stating now they know what we deal with. We eventually took pride in the fact that we had the scariest Drill Sergeant in Charlie Company.

In the bay on Sunday when we finished cleaning, guys were able to shoot the shit and unwind while we waited for the religious services to end. Sundays were beginning to be a huge reprieve for many of us. Even if I had to clean, I didn't mind one bit, it allowed me to be away from the rest of the company. Love them or hate them or both, I took whatever solace I could get. And it was rare. I would usually be sitting by myself at the window, writing a letter or working on this journal and gazing out at the free world. Occasionally I would stare out the window at the world, seeing the green trees and picture driving around with friends, windows down and music up on a beautiful warm, sunny day. I imagined myself running or working out on my own, sleeping in, and enjoying the beautiful weather back in California. I would at times be joined by my close buddies Chrizzle, a guy named Taylor that Drill Sergeant Price nicknamed "Ranger Cook," or my friend Jake since he and I discussed numerous subjects in the civilian world like music, travel, and food. Chrizzle introduced me to a techno-like music genre called Dubstep, and we got talking about video games and kick-ass movies. Later on, he and I made a list of songs off of memory that we were dying to listen to, and an exceptionally long list of food we missed dearly being subjected to good ol' Army chow. It wasn't *bad* chow, we were just deprived of all the sugary and salty foods and treats we would've been grateful to have at that moment.

Every time I would walk through the bay on a Sunday, I would pick up a broad range of conversations from family to clubbing, finances to sports. It was always interesting to hear who was talking about what, but as soon as it got to negatively talking about this girl or that girl, or this guy and that guy in the platoon or company, I walked away or my thoughts dwelled elsewhere. I dislike talking about other people behind their backs, even if I don't like them. If I don't like you, I'm just not going to talk to you period (unless I need to talk to you for professional reasons, I'll still be an adult about it), not go behind your back and spread shit to make myself feel better. That's high school stuff. I hate drama, but little did I know at the time there are many people in this world who live by it, and some can't even breathe without taking a daily trip to the rumor mill. **<sarcasm>** I want to grow up to be *just* like you **</sarcasm>**.

One platoon member (we'll call him "Ether") discussed business and the best types of accounts and IRAs. I listened to him on occasion to get some good pointers; he stated the best IRA is a Roth IRA. A fun discussion was when I would catch guys talking about females in our company. Going over which ones they found attractive, which ones they would bed, blah blah blah. And we all knew the females were talking the same about us, among the certain looks, notes being passed, and conversations in the field. Everybody's trying to get their game on. Hell, no sex, masturbation, porn or embracing the opposite sex for three whole months, nearing the end of basic, *I* was eyeing a few females myself. Discussions that were frequent in the bay consisted of setting priorities when returning to the civilian world, and planning what to do with the money we were making. Many guys said they would pay off a debt (myself included), some were going to invest, and others were going to put their hard-earned money in savings, or toward their families or their vehicles. And hard-earned money it was.

3rd Platoon members in my bay would come up with interesting and effective new workouts for personal pt. One such workout was the IBA burnout; I and a couple other guys donned our IBA vests and someone put another IBA on our back while we were all in the front leaning rest, so we had upwards of 40 pounds on us. For one minute, you perform one push-up every fifteen seconds then hold the up position (also called the high plank). When the final fifteen seconds hits, you all "burnout," performing as many push-ups as you can. One of my favorite exercises was the mid plank---forming a straight plank with your body, you rest on your forearms with your hands together. A few battle buddies would watch as I held it for eight minutes, twenty seconds. Stewie would time me every time. As a kid from 1st platoon watched and waited 'til I finished, he decided to try it out. I rested while watching him hold the mid plank and Stewie timed him. He beat my record by two minutes. Now I had to beat ten minutes. It is said if you can hold a plank for two minutes, you have a strong core. We were way beyond that mark!

Stewie timed me on the plank again another day and I reached a crazy 15 minute, 48 second hold. My body was shaking and I was sweating, my arms, abs, and legs burning. I didn't even drop suddenly when I couldn't hold it anymore, I slowly sank to the floor…and damn did it feel good to finally rest.

During White Phase, male and female recruits were receiving "Dear John/Dear Jane" letters from their boyfriends, girlfriends, husbands, wives. It's only been a little over a month and already these people's mates are high-tailing it out of the relationship, and for some, out of their place with all of the recruit's stuff. One big issue is impatience, another is sex. The partners don't want to wait three months for their mate to get back home whether they can't deal with them being gone for so long, or they can't stand not having sex or sexual favors for an extended period of time. I've been told this one guy's wife took their child and a bunch of the stuff in the house, leaving the guy with little-to-nothing. I was told by a female buddy that her boyfriend left her because he couldn't deal with the distance. I'm not judging, but it really sounded like he wanted more tail and didn't want to be "that guy" to cheat on his girlfriend. Who am I to judge, unless it's told to me, I shouldn't assume what goes on in other people's lives. I have read and experienced vicariously through others the exact situation I am talking about, though, so there is some truth to it.

I have griped about this to others before, the military's not for everyone. Even more so, for the ones who join and have partners, the civilian partner has no fuckin' idea how the military is run or the shit we deal with---unless they're accustomed to it through having a military family. Unless they are a loyal, loving, caring, and sympathetic partner who wants to be 100% supportive of their military man or woman, all they will care about is themselves and the money that is being made by the one person trying to support them, earned by the person who is trying to better themselves and their lives, and the lives of their families. Here I underline the fact that there are Americans who have their priorities out of line. Many people exploit the fact their mate is in the military for personal gain or status. I hate to say it of the very people we fight for and protect freedoms for, but it is a sad, cold fact. Much of life is unfair and dark and chaotic, and I may be exaggerating a bit, knowing there are more pressing matters in this world but this is one issue that needs to be addressed, in addition to veteran suicides.

Since the United States military has become an all-volunteer force, our numbers have dwindled since the Korean War. The United States fighting forces make up less than 1% of America's population. We are the .45%, go through hell in many instances, and don't get paid a great deal (contrary to popular belief). There are people who look down on us, think we are not needed, and even state we have too many benefits! The men and women of the military signed a contract; they signed a check that is one day possibly paid for with their lives---the ultimate sacrifice. They signed and they volunteered. Granted, not all personnel in the service are saints, but good or bad, we fight for the good of America, no matter how bad it may be at times. If you don't want to stand behind us, stand in front of us. Not a whole lot of people put the proverbial shoes on their feet to walk the walk. Not many people can, or they simply choose not to…and that's fine. That's where we come in. No matter how bad you shit-talk us, no matter how much you hate us in peace or war, we are the ones going to battle. The service members are the ones who fight the monsters; some even become monsters to fight them. Loved or hated, we are the sheepdogs that protect the sheep from the wolves who want to slaughter the herd. We embrace the suck so you don't have to. It's not that we can and others can't (although there are some people out there who legitimately can't), it's that we will and others won't. Again, we signed the dotted line and volunteered. So next time someone complains that the barista got their coffee order wrong, or they get anxious because they have nothing to wear on a Friday night, just remember what we go through, and that the military doesn't have much of an abundance of choices of how to spend their downtime while overseas. First world problems, I guess.

I grew to find solace on Sundays. When everyone went to religious services, I would wash my bay's clothes. Sounds like a shitty detail, right? Wrong. Three by three, the people next on the list would bring their laundry bag down to the laundry room and whoever's was done washing and drying, I'd put their bags aside and have them come and pick them up. It was peaceful in the laundry room; I could do a little pt, talk to a few buddies here and there (it was either someone who helped me with the list or someone that came to drop off or pick up), or work on writing this book. My system of making a list, sending three down to wash, then mark off the next one(s) up was so well-done and in such a fast manner, that whenever Sunday came around, people would try and be the first ones on the list. The last two Sundays of basic I handed the job to three others that asked to take it over. It was fun while it lasted and I enjoyed the discussions that were had.

Throughout White Phase, with all the FTX's, getting smoked, working as a team, and growing accustomed to the tedious and demanding schedule, I began to see and experience what a soldier goes through. It's not easy; hard but fair. To hear what soldiers do and to watch movies and documentaries is one thing. Experiencing it all is another world entirely. Now I have two views, being on the other side of the fence. Much of a civilian's point of view is what is seen on tv, heard from a family member or friend that's current or former service, or what is described in the news. A little trivia: the word 'news' stands for North, East, West, South, which aptly encompasses all directions of media coverage. Back to the main essay, it is easy to say, "I can do that," or "That sure seems pretty tough," but actually *doing* it proves you either can or you can't.

The act of doing—*that* is experiencing. Pushing your mind and body to its physical and mental limits, forcing yourself forward, pushing through the pain and exhaustion, then passing your limits and working on autopilot is a peculiar feeling, but one I feel is necessary for people to experience what they are capable of. Muscle failure helps break your physical barriers and builds you up to go farther. Push-ups. Sit-ups. Flutterkicks. V-ups. Until you can't go anymore. 8-count push-ups. Leg tuck and twists. Supine bicycles. Until your muscles start to fail and your body decides to stop doing pt. The next time you reach muscle failure, and the times following, you'll realize you can endure more and work out longer before you start to fatigue.

After breakfast one day, all platoons are sent to their platoon bays (as opposed to sleep bays where it's same gender). We sit around the bay as our Drill Sergeants hand out peer-to-peer evaluations. Take one, pass on the rest. I was a little surprised when a handful of my platoon, both males and females, wrote me down as one of the top three soldiers in the Wolf Pack. I mean, I'm aiming to be *my* best and bring out the best in others, work well as a team, and so on. But to discover that there are a number of people who look up to you and list you as one of the top three soldiers in your platoon is not just an honor, it's a feeling I can only describe as being related to a group of people who are grateful for helping them reach a goal in their lives. Not a very good analogy, but since I don't have children, I can't relate it to a younger generation looking up to me…although there were several buddies who were younger than me.

The Modern Army Combatives Program (MACP) is a practical hand-to-hand training program that features jiu-jitsu in the beginning stages, then increases in intensity with reversals, counters, and disarms in the advanced levels. There were one-on-one matches held and the Drill Sergeants even joined in. It was a blast watching soldiers from different platoons roll with the Drill Sergeants, testing their skill or what they recently learned.

My company held one of our record APFTs (Army Physical Fitness Test) early in the morning, before the ass-crack of dawn. It was freezing and many of us were worried we wouldn't knock out enough push-ups or run fast enough to pass our 2-mile due to the cold conditions. When we started our 2-mile run, by the second lap my legs and lungs were screaming at me. They were cold, not warmed up, and I wanted to stop running…but I couldn't or else I would fail. I'll always remember what Drill Sergeant Price told the platoon.

"Why do I keep running? So I can stop running. I don't let up and I don't stop until I finish."

I kept on keepin' on and by the fifth lap around the track my body finally warmed up and I felt better. I passed the PT test, but it could've been better, there is always room for improvement. Upon finishing our pt test, we showered then marched to breakfast, everybody uplifted and feeling like a weight was lifted off our shoulders. Not everyone passed, but those that did celebrated with a donut or cake, whatever sugary delicacy was being served in the chow hall. The funny thing is we paid for it after, the Drill Sergeants made sure of that. They were the ones laughing at us figuratively and literally, both while we were eating the sweets and while we got smoked.

Charlie Company marched out to one of the ranges for rifle qual. The sky looked gloomy and a little while after we arrived it started raining, but as it is said in the Army, "If it ain't rainin', it ain't trainin'." Concurrently, as recruits did their best to qualify, the rest of us practiced compass and map reading while we waited. When our rotation was up to qualify, the rain didn't let up at all and the mud and puddles increased, making the act of qualifying a bit challenging. I approached my lane, set up my positioning and laid prone, awaiting the first set of targets. I was wet, cold, and muddy, but that wasn't going to stop me from qualifying. Range Control wasn't immediately going to decide, "Alright, Mother Nature's PMSing, let's hold off on finishing this." The enemy does not wait for adversaries to be perfectly fed or get 7 – 9 hours of sleep, and the enemy definitely doesn't hold off on attacking simply because there's "a storm brewing." Terrorists, criminals, and death are not considerate or honorable. My iteration finished, collected our empty magazines and then headed over to the Drill Sergeants training recruits on land nav, and the remaining recruits still on the range retried meeting at least the minimum standard for qualifying.

The rain and thunder were welcomed that day, whether or not I was qualifying in it. I loved rain and it put my mind at ease; watching it fall and watching the lightning in the distance, counting how many seconds it took to hear the thunder to know how close or how far it was. The last few minutes before we were called to roll out, I stared off into the distance and wondered what my friends were doing as I was in a previously unvisited state training to become a soldier.

In reference to recruits wanting to get by and not be noticed by Drill Sergeants, I got noticed indefinitely when the platoon went back up to the bay after breakfast one fine morning. Our infantry Drill Sergeant started yelling at us about securing our lockers, and he walked right up to me and told me to get up. I did so, stood at parade rest, and he asked if I secured my locker.

"Yes Drill Sergeant."

He asked me again, "Are you sure Private?"

Again I affirmed, "Yes Drill Sergeant, I make sure my locker is secure before going anywhere."

He then held up a dial combo lock and asked, "Are you fucking sure, because this looks like your combo lock. Did you lock your locker this morning?"

I looked at the lock then back at the Drill Sergeant. I looked him right in the eyes and said, "Yes, I am sure, Drill Sergeant."

He got real close, almost nose to nose, like when a guy looks hard at another guy, really egging him on to fight.

"Are you fuckin' lying to me, Private?!"

Perplexed and feeling anger rise up inside because I hate when I'm being called a liar when I am *damn sure* I know I'm not lying, I replied, "Negative Drill Sergeant! I *make sure* I secure my locker every time before leaving the bay."

"Then tell me why the FUCK am I holding your combo lock, Private!?"

Wondering if he was messing with me or if it was really mine, I answered, "I have no idea Drill Sergeant!"

And with that, he chucked the lock against the wall at the front of the bay. I went over to inspect it; I picked it up and looked it over, then walked over to my locker. My locker had no lock on it, so it had to be mine; looking over the combo lock again, I noticed it wasn't cut. I opened my personal effects drawer and didn't see a combo lock anywhere. So this one *was* mine! How the hell did this lock come off without being cut? The only explanation was that I didn't clamp it shut after putting it on my locker. I guess that tiny detail fell through the cracks when I had other thoughts and worries on my mind...but it should've been second nature to clamp the lock secure. This is why it is hammered into us about having a strict attention to detail.

There was one Drill Sergeant in Charlie Company we nicknamed DJ Wyma. For me, his smoking sessions were a love/hate thing; I hated them because we would be smoked for what seemed like forever and the exercises he made burned real bad, but I loved them because of how he presented them. He smoked recruits and treated it like it was a game, realizing a bunch of us loved to play games. "So you want to play a game?" He always asked, followed by us getting in the front leaning rest or on our backs. One smoking he gave us was particularly my favorite; I didn't care how bad it burned or how tired I got. Drill Sergeant Wyma would command us almost like singing a cadence, and we would echo the second part while we were getting smoked in the leaning rest.

DS Wyma: "All the way down and hold it."

Recruits: "And hold it!"

DS Wyma: "And hold it."

Recruits: "And hold it!"

DS Wyma: "All the way up and hold it."

Recruits: "And hold it!"

DS Wyma: "And hold it."

Recruits: "And hold it!"

DS Wyma: "Half-way down and hold it."

Recruits: "And hold it!"

DS Wyma: "And hold it."

Recruits: "And hold it!"

This would go on long enough where people's arms were tiring and some flopped to the ground. Others were shaking, trying as hard as they could to hold themselves up so the Drill Sergeant wouldn't see they were resting and force us to get smoked longer.

"Regard your soldiers as your children, and they will follow you into the deepest valleys. Look on them as your own beloved sons, and they will stand by you even unto death!"

-Sun Tzu

"The Art of War"

Every now and again some of the Drill Sergeants gave "pep talks" and personal, motivational speeches to the platoons which felt very much like a parent talking to their kids. It really was motivational, and after days or weeks of smokings, yelling, and training, a sit-down with a Drill Sergeant felt fantastic----not because of the ability to rest during the duration, but for the reason that the sit-downs felt personal and it was sort of a bonding experience. These sessions were essentially called "hugging" the platoon. One NCO who was going to Drill Sergeant School after our cycle was with the company on many occasions, on training missions in the field with us, around the company area, even being with us during the range or other events. I liked him a lot because he had the same communications MOS I was training for. And he was really cool and down to earth. This NCO, Staff Sergeant Lace, gave our platoon a few of these lectures and talks, but I remember he was pissed during one unforgettable sit-down.

"I've talked to you guys long enough, Third. Many pointless talks with you guys---do you want to know why they're pointless? Because you guys keep getting in fucking trouble. You guys keep fucking getting smoked. I think this'll be the last time I talk to this platoon, 'cause I'm wasting my breath."

The scolding and smoking the Wolf Pack endured were mostly thanks to Turtle. The whole bay was silent, each of us either looking around at one another or looking straight down, introspective and slightly brooding, somber expressions all.

Staff Sergeant Lace continued, "You know, there are only six people in this whole platoon I would take with me on deployment. Six."

He paused for a moment then said, "Scratch that, seven. You too, Askedall."

I didn't bat an eyelid but I did look down, like a modest, "I don't deserve that…but does he mean it?"

That got my hopes up. It was a morale booster for me, bringing back all the reasons I joined and once again encouraging in me the confidence, loyalty, and honor to do my best and graduate. What Staff Sergeant Lace

said just then stuck with me. That was the recognition I have been

looking for since childhood. The compliments on my drawings, the

"attaboys" for good grades, and the thanks I received over the years from

my parents…none of it compared to that moment. If there was one time

in my life so far that ever filled me with great pride and self-satisfaction-

--that was it. To be trusted with the life of another individual overseas in

a war zone was an enormous honor and responsibility, a responsibility I

wouldn't be so quick to take up, but one I hope and pray would prove me

the right man for the job at the moment of truth.

"Greater love hath no man than this,

That a man lay down his life for his friends."

John 15:13

As the weeks went on, mail call started to become fun. I mentioned

earlier about the number of push-ups each person had to perform for an envelope and a package, and some people received five or six envelopes in one sitting. The one Drill Sergeant the Wolf Pack liked having mail passed out by was Drill Sergeant Price. Each time someone got mail, he would tell them to "beat your face," meaning do push-ups. This eventually became a "game show" of sorts, and here and there Drill Sergeant Price would yell, "Come on down! You're the next contestant on..." and the whole platoon would yell in sync with him, "Beat Your Face!"

At 2200 the females would return to their bay, the lights were turned off, and some guys would sleep, some would stay up writing letters, and whoever had fire guard would stay up until the shift change.

The Wolf Pack started with 53 recruits and through White Phase we lost a handful, and three we lost not even three weeks into Red Phase. Most of the recruits were sent home due to injuries they received or because of family issues. The guy I met at the airport on the way to Fort Jackson pleaded to be sent home, he missed his wife and kid too much. Uncle Sam sent him home before reaching White Phase. At best, that's four weeks. Imagine being on a deployment for nine to twelve months not being able to see your family. Then again, I'm not a father. A buddy of

mine, Martinez, was sent home to help her hip recover, and she was told

she could come back to try again. Two individuals were recycled back to

Red Phase to (if they wanted) complete basic with another company.

Section IV: "No smiling until you graduate" –– Blue Phase of BCT

Basic training felt like a dream for the longest time, and the company was spreading around a little countdown that grounded all of us.

55 days and a wake-up…

30 days and a wake-up…

21 days and a wake-up…

We all felt we were dreaming due to various reasons; the training, only wearing two uniforms for three months, the constant exhaustion, and no computers, phones, or mp3 players. Being here, doing this was completely different from who we were and what we did outside the Army. Now that we finally made it to the last phase of basic, the company's remaining recruits were excited to have come this far and almost be "free."

One day the company loaded a line of buses and was driven to the U.S. Heavy Weapons range. Hooah, now for the big guns!

M2

The first one I tried out was the M2, or "Ma Deuce." She was the fat lady that could sing, stopping any SNAFU (Situation Normal, All Fucked Up) dead in its tracks. Upon finishing the Drill Sergeant's demonstration of how to use it, I sat down and got into position, thumbs on the trigger when I was ready---and I was ready. I held the trigger down for two seconds and a loud, heavy **BOOM-BOOM-BOOM** left the barrel, and my time on the M2 was over. I was craving more, like a virgin who got over the awkwardness of the first time and wanted to better himself. The power and lethality of that weapon were awesome. When fired, the .50 caliber bullets explode from the barrel at approximately 3,050 feet per second. The ground under the barrel was blackened by the heat and every time the weapon was fired, the rocks under it would jump.

M249 SAW (Squad Automatic Weapon)

The SAW machine gun is one of my favorites of the heavy weapons, albeit slightly cumbersome lugging around. I love hearing the *rat-tat-tat-tat* as you assault your target, and feeling the recoil is a little like the vibration of an Xbox controller, but two times bigger. Just don't fire blanks with the M249, it *always* jams with blanks and then you have to clean up the crud it leaves behind...but that's like firing blanks in any weapon.

M240B machine gun

The 240 is a beast. If you think the SAW is heavy, don't complain about it to someone carrying this bad boy. Weighed at 27.6 pounds and firing 7.62mm rounds, the "Bravo" is an excellent weapon to deter enemies from advancing.

M203 grenade launcher

The M203 fires a 40 mike-mike grenade, and launching a grenade from an M203 is almost exactly like how it is in video games, minus the crosshair reticle. You raise the M203 attachment to the height you want to lob it (depending on the distance to target) looking through the leaf sight, you pull the trigger, and with a *thump* the projectile arcs its way to the destination, exploding on impact. The Drill Sergeant gave us two tries to hit our target, and I missed the first, but second was a direct hit. I slid open the tube and ejected the empty casing. This ain't the movies and video games suckers, this is for reals.

M136 AT-4

The AT-4 is an anti-tank/anti-armor rocket launcher that is extremely light. Only the platoon guide (PG) and assistant platoon guide (APG) of 3rd platoon were able to fire live rounds, everyone else fired dummy rounds. Again, just like the M203, we all had two chances to hit a target. It seriously was not difficult to hit a target; you line up the front and rear sight posts on what you want to shoot, and you push the button (the lack of a handle and trigger surprised me a bit). When the PG and APG fired the live rockets from the AT-4, the boom was loud and thunderous. I would never want to be on the receiving end of that thing.

We went to the rifle range one final time to qualify. While waiting for our rotation to shoot, I and the rest of the company got down in the front leaning rest because some of the company wouldn't move with a purpose from one location to another where Drill Sergeant Price was waiting, but a select few recruits still didn't get the message when one of the other

Drill Sergeants told them to stop talking. We were all in full battle rattle, attempting to hold ourselves up in a high plank while Drill Sergeant Price talked to us. Do you know how difficult it is to hold yourself up for an extended amount of time with a flak vest and helmet on, not including an assault pack if you were unlucky to be wearing one as you got smoked?

"Why are you down in the front leaning rest, Pri'ates? Can someone answer me that?"

No one answered; half thinking the question was rhetorical, the other half thinking that answering the question may lead to a lengthy smoking.

Drill Sergeant Price asked again. "Does anybody know why you are all getting smoked?"

I decided to speak up. "No discipline, Drill Sergeant."

The Drill Sergeant looked around, searching for the goober that opened his damn mouth.

"Who said that? Stand up Pri'ate."

I got up and stood at parade rest.

"Say that again Pri'ate."

"We're in the front leaning rest because of no discipline, Drill Sergeant,"
I responded.

"That's right, Pri'ate. You are all down there because of no discipline."
He then turns to me and says, "Go to the bleachers, Pri'ate," releasing me
from the group smoking session.

"Drill Sergeant," I start, "Permission to get down and join my company
in punishment?"

"Negative Pri'ate, get outta here."

I tried. I picked up my rifle and ran to the bleachers where the other
recruits were waiting to shoot.

Flat, Dark Earth

After the heavy weapons range, Charlie Company hopped in the back of
a couple trucks we call a 5-ton, and group after group, we were
transported to a wooded location where we downloaded our rucksacks
and awaited our next order. When the majority of the company was in

the area, the Drill Sergeants relayed that we were to find a spot, set up our sleeping bags for the night, and be ready to move before the sun came up. Pitching our tents would take too long and that would be time we wouldn't have to pack them up in the early morning. So we rolled out our sleeping bags onto our PT mats and got comfortable. I used my ruck as a pillow and got intimate with my M16 again, hugging it in my sleeping bag. Lying there on the cold ground underneath the blanket of stars brought me back to my homeless days, except now I was getting paid to sleep on the cold ground, and I had a more substantial cold weather sleep system than I did when I was younger and on the street. I awoke to rustling and movement in the darkness. The company was up and starting to pack, flashlights dancing around the woods, some white, some red. I quickly got up, shook my boots upside down so I wouldn't have an insect or arachnid surprising me, and stuffed my ruck.

We humped it to our new campground and after we settled in, the company marched over to a MOUT site (Mobile Operations in Urban Terrain). This was going to be different from our typical room clearing practices in almost every way; there were going to be unit bigwigs spectating, a camera crew filming for the basic training dvd, and a rotation of one platoon as the good guys and the other three as the opposing force, or opfor. This exercise was going to be huge. The good team was told they have the ability to use smoke grenades and frag grenade simulators (basically sims that have a big, loud bang). There were platoons ahead of 3rd, so as we waited we picked squad positions and worked on our floor plan in the dirt, carving buildings and setting rocks and twigs for platoon positions and lanes of fire. Our objective was to rescue a civilian hostage in a very hostile village and exfiltrate with the hostage alive. Knowing that the other platoons envied us for being the best platoon in Charlie Company, coupled with knowing the Drill Sergeants from every other platoon hated us, we knew we were in for some heavy shit. The opfor and the Drill Sergeants would not make this mission easy. We had on lasertag-like harnesses and mounted lasers on our rifles, and the harnesses would beep if hit, simulating being shot. It wasn't as accurate as paint rounds making a mark, but it was a helluva lot cleaner.

My Black Hawk Down Moment

Finally it was the Wolf Pack's turn. I was nervous because one, I was

Alpha Team Leader of my squad and wanted to do a good job and not

fuck up, and two, I wanted the Wolf Pack to succeed. I'm sure the rest of

the Pack shared my dread. We started the mission slowly patrolling

through a grassy field, the village lying in wait up ahead. All four squads

separated so if a notional grenade came our way, the whole platoon

wouldn't go down. My heart started racing as we got closer. Suddenly, a

mortar sim went off (loud, sharp whistle followed by a boom) and we hit

the ground yelling "INCOMING!"

Three seconds later we were getting up when shots were being fired at

us. Every squad ran to the nearest building, most of us firing back at the

opfor. It seemed to me they were in every window, every doorway. Just

like the times I went paintballing with friends and my younger brother,

adrenaline surged as we assaulted forward at full speed. Nobody wanted

to "die." The Drill Sergeants picked off two to lie down and be casualties

as their laser harnesses beeped. Two squad members dragged them to safety as our squad leader commanded a soldier post security on every corner of the building.

"Man down!"

Dammit! I thought. *We have to find this hostage and get out of here.* An arty (artillery) sim went off. The squad leader ordered two squads upstairs on the other side of the building. The radio chatter was almost constant, and that was good, it meant communication was being made and situational awareness was being shared. I was kneeling by the door on the back side of the building, head on a swivel. I then spotted a bad guy on the second floor of the building further down, aiming our way. My buddy Endicott was taken out by the sniper, his vest beeping as he lied down, pissed off he was eliminated from the exercise. I immediately shot the guy and kept looking around, but mostly focused on the door, windows near me, and the door and windows of the building where I spotted the sniper.

More booms, more yelling, and more shots fired.

Our APG, Wyche, was with me and it was only us two. The rest of the platoon was pulling security, clearing rooms, or "dead." Wyche's radio burst with more transmissions.

"Echo 1, this is Echo 2. Second floor clear. We have the hostage."

"Roger Echo 2. Echo 1 coming up."

As 1st and 2nd squads were meeting upstairs, Wyche got orders from 1st squad leader to clear the ground floor so they can leave through the back door, in case there were any opfor waiting at the front. The only squad member Wyche had to assist him in clearing the ground floor was me.

"Alright Askedall, we gotta clear this floor so they can bring down the hostage. It's only you and me."

"Roger that, let's do it," I replied.

"Okay, you go in first."

Do I have to? I thought.

Wyche got behind me as we stacked on the door. I patted his arm signifying I was ready, and then he returned the gesture.

Here we go.

I charged in, cut my section of the pie and went around a wall to the front. Slow is smooth, smooth is fast. I took a knee in the corner opposite the front door, watching all available spaces.

"Clear!" I yelled.

"Clear!" yelled back Wyche. He got on the radio and relayed that the ground floor was clear, ready for extraction. One of the infantry Drill Sergeants from 2nd platoon peeked through the window. He looked curious.

"Clearing rooms with just two?"

"We've got no one else, Drill Sergeant," I answered. I think his question was half rhetorical.

"I didn't say anything." And with that, he continued to monitor the exercise.

"Coming down!" shouted 1st squad leader. As soon as 1st and 2nd squads made their way out the door, Wyche summoned me to regroup.

"Askedall!"

"Moving!"

Endicott lay outside where we left him momentarily. Wyche wanted to cover our escape while we dragged Endicott to the rendezvous point, so he asked Drill Sergeant Price to pop smoke.

"Where do you want it?" Drill Sergeant Price asked.

Wyche pointed. "Right there, Drill Sergeant."

The Drill Sergeant pulled the pin and lobbed the smoke grenade, yellow smoke billowing. I took Endicott's M16 and slung it over my shoulder. Wyche and I each took one side of Endicott's laser vest, nodded to each other, and started pulling our notionally fallen buddy. It was a long way to the rendezvous, but our platoon was waiting for us and firing at various points in the village, covering our exfil. Once we reached the rest of the team, Wyche and I posted up on the wall and joined everyone else in the wall of bullets we were sending downrange. I had to reload so as I got down below the open window, Wyche's blank casings were ejecting in my direction, and one landed under my shirt and started to burn my neck. I rushed frantically to get it out and get back to reloading (but the reason I was frantic was because it was burning the shit out of my neck). The Drill Sergeants then yelled "endex" to end the exercise, not because of my insignificant incident, but for the Wolf Pack's completion of the mission.

The platoon followed the Drill Sergeants to an area outside the MOUT site to go over an after action review, all of us unloading magazines and chambers and putting our rifles on safe while we walked. The Drill Sergeants cussed and scolded us for all the mistakes and told us how and what we could have done better. Then they ended with our high points and congratulated us on getting the hostage out alive and not having the entire platoon get killed, like what happened with one of the platoons before us. Fuckin' A, Wolf Pack. We got it done.

*   *   *

That night the company had another mock mission---the Night Infiltration Course. The sun was long gone and it grew cold real quick. Again, the platoons were going in order, so as 1st and 2nd platoons finished, 3rd and 4th were up. This was a joint operation which included two platoons in one iteration. At the start, we trudged through a sort of irrigation ditch, water almost up to our knees, submerging our boots. This wasn't even a worry in my mind, thinking of the task at hand and focusing all my natural night vision on my team and the surrounding area. We climbed up a sloped concrete wall to get out of the ditch, and right when we made it up and out of the trench, a .50 cal M2 began firing over our heads. Everyone dropped and started crawling through the dirt, and continued crawling when we arrived at what seemed like a field of barbed wire. There was shouting all around me, and I was shouting at my team to keep moving forward. I could tell the two .50 cals were firing live ammunition because I could see the tracer rounds zip through the night's sky. There was abundant shouting and a constant **BOOM-BOOM-BOOM**. I could hear the Drill Sergeants scream at a few people to keep crawling.

When the two platoons made it completely through the barbed wire, the shooting stopped. We all stood up and crept forward, rifles at the ready. All of a sudden, there were silhouettes ahead of us and we opened fire with blanks. Both platoons moved forward and the Drill Sergeants ended the exercise.

On the march back to the campsite the temperature was dropping and I could definitely feel my wet, soggy socks now. Right when I got to my tent, I started changing into PTs to sleep in. It took me a moment to remove my boots; I was shivering, the laces had to be loosened, and it was tough pulling them off resulting from the wet socks. The whole time I was changing I kept thinking, *I have to keep moving, have to stay warm, I have to stay moving to stay alive, the sooner I change the sooner I can get into my sleeping bag and get warm.*

*   *   *

On the last day of the FTX, Drill Sergeant Nelson punished all of 3rd platoon for one member's lack of purposeful moving from point A to point B. Our smoking was low crawling—the worst and slowest of the individual movement techniques—out to 50 meters, then crawling back, and doing it again. This wouldn't have been so bad if we weren't crawling through clay. The red earth got all over our uniforms, helmets and vests, and in the cracks and holes of our rifles, and I assumed when we eventually got to downloading and cleaning our gear the clay would leave a red tinge on everything. While we're slow crawling, the whole time Drill Sergeant Nelson was yelling at us, educating us as to why we need to be hasty in moving from one point to another, and stating how important teamwork is. Slowly moving through the clay, I bring back a thought about blue phase; just because it's the final phase of basic, does not mean we get off scot-free. The day was beautiful and a very cool temperature, and I started sweating. The clay was sticking to my face and I kept thinking of this FTX and the end of basic. We were in the homestretch and no matter what was thrown at us, I enjoyed this basic soldiering and drilling.

Intimate with the Sand Pit

Back at the company area, we marched to the central issuing facility where every recruit went to be issued gear and uniforms. We were receiving our Class A uniforms (the formal Army uniform that holds medals and ribbons) and berets. The wait-and-receive process started out quiet, until much of the company began talking amongst each other. The noise grew louder and louder, until one of the Drill Sergeants yelled for everyone to shut up. The silence lasted only a few moments then built up again to a clusterfuck. Again a Drill Sergeant told everyone to lock it up, but this time it was Drill Sergeant Price, and he warned the company for the last time. I was talking too but at a low tone, it was many others getting rambunctious. The whole company was going to pay for their inconsideration and ignorance. I was getting pissed but cooled down and said to another recruit, "It is what it is, some people never learn. But it's okay; they'll get what's coming to them. It's just stupid we all have to suffer."

The recruit agreed and understood exactly.

The second and last time, Drill Sergeant Price told everyone to lock it up…but this being the company's third strike, Drill Sergeant Price stated that we were all going to get intimate with the Sand Pit. *Shit*, I thought. We returned to the company area, duffels on our backs, and put everything away in our lockers. Drill Sergeant Price was walking through the bays, telling every platoon to get out our gear and to be in full battle rattle. Some of us were actually hoping he either forgot or was joking about the Sand Pit. Now *that's* funny---a Drill Sergeant forgetting. And the way they joke is through the expense of others to teach a lesson, so I guess this was going to be extremely funny. Charlie Company marched, full battle rattle and rifles in hand, to the big rectangular pit of sand where we would learn the lesson *again* never to ignore or disregard a Drill Sergeant's order.

"Line up in four ranks, Pri'ates."

The company created four ranks facing Drill Sergeant Price.

"Right; FACE." We executed a right face, now facing the Sand Pit in four long lines.

"Alright Pri'ates, the first drill is going to be the High Crawl! And when you get to the other side, you're going to continue back across with individual movement drills. Back and forth."

I watched everyone ahead of me high crawl through the sand. After I made it to the other side, I got up, fell in behind another battle buddy, and then got back in the sand when my turn came around again. Another drill going back, another drill going across, and so on, back and forth. Recruits were losing their magazines in the pit, and by the time we finished our smoking, we were sweaty with sand everywhere---on our faces, in our noses, hair, in our pockets, boots, in our shirts and underwear, and in our rifles which we would be cleaning before lights out, as per the Drill Sergeants' request. 3rd bay knocked as much sand as we could off and out of everything before entering the bay, but it wasn't enough and sweeping the bay was another task we carried out before lights out. Of course we didn't have ample time to clean rifles *and* the

bay before lights out, so some of us were still up brushing away and
wiping down every piece of the rifle before our heads hit the pillow.

*   *   *

Later that week, we went to the grenade bay, where we practice lobbing
simulation grenades a safe distance away before we could go on to the
live grenade bay. We had two chances to throw the sim grenade outside
the danger-close radius, and if you screwed up both times, you would get
a CW written on the side of your helmet and you wouldn't be allowed to
throw the real, live grenades. Since CW stood for "chicken wing" and we
really wanted to throw live grenades, me and a good majority made it a
goal to chuck those sims as far as we could. I passed the test and
continued on to the next level. You feel the shockwave even through the
concrete barrier. An explosion of a live M67 frag grenade is almost
exactly like the explosion of a frag grenade in the video game Halo; big
burst of shrapnel with a cloud of smoke. It's not like I looked at it after I
threw it; I watched an explosion of one through a reinforced window, far
from the barrier where the grenades were thrown.

One night the 1st platoon infantry Drill Sergeant snuck into 3rd bay unbeknownst to us. He is a short infantry sniper who can enter and exit so silently and without a trace, and he did so that night. He discovered that our bay had absolutely no one on fire guard, probably due to the sheer exhaustion from the days previous because of training. Although there has been a time when I woke up late for my shift, that night I wasn't on shift but the ones that were said they were up for their shift and we were all puzzled. It just so happened that the Drill Sergeant came into the bay when that specific shift was sleeping. *Just our luck*, I thought, shaking my head.

Our punishment the next night was from our infantry Drill Sergeant of 3rd, who was told by the Drill Sergeant who made the discovery. Our punishment consisted of the whole bay doing guard duty, roaming the bay in full battle rattle and ACUs, in fifteen minute shift intervals. Each shift had fifteen minutes to get into ACUs and all our tactical gear and roam the bay, rifle in hand. And it's not like we could skip out on the ACUs and boots and don our gear over our PTs; our Drill Sergeant not only said specifically not to do that, but he would also stop in the bay periodically to check on us. So getting out of PTs, into ACUs and boots, and putting on our gear to walk around the bay with our rifles, all within fifteen minutes? Fuuuuck……..

We would be spending much of that time getting into and out of gear and uniform, and by the time the shift after yours ended, you'd have to get it all back on again. It was frustrating for all of us. A few of us were quietly fuming, waiting for this smoking/fire guard to be done with. It was around 0100 when our Drill Sergeant came up to the bay for the third time since 2200 and saw the bay was diligently carrying out his orders. He told us we could stop and return to our original fire guard on rotating shifts, back to the way it was supposed to be.

"You're not going to fall asleep on guard duty anymore, are you?" he asked.

"No, Drill Sergeant," the bay answered.

"Good. Have a good night, Privates." And with that, the bay was relieved we would be getting some much needed, much wanted sleep.

During whatever free time the company had after our berets and Class A's were issued, we worked on forming our berets. First we shaved them with razors to get that furry, fuzzy look out, and then we dampened them

with water and put them on, forming them and pulling them to feel good

and look better. Some of the guys and girls weren't doing so great

because their berets looked like pizza hats, big and plumed like a hat of a

chef, it was pretty funny. The Drill Sergeants had a good laugh and made

those recruits keep working on them.

Closer and closer Graduation Day crept. The weather was getting

warmer and we were doing more tasks outside, like washing the 5-ton

truck and picking rocks, stones and pebbles out of the grass and putting

them with the others by the formation area. In the middle of one assigned

duty day, a handful of us males from 3rd ran up to our bay, and on the top

floor, across from 3rd bay, were two Drill Sergeants from 4th platoon right

outside 4th platoon's bay. They ordered us over and we obliged. They

told us to toss the whole 4th bay. Anything and everything in its proper

place, we were to fuck up. Slightly excited, I and the others of my

platoon ran through the bay, knocking over lockers, tossing bunks,

pulling sheets and blankets off of bunks and throwing them all over the

bay. Boots and pt sneakers were thrown everywhere, mixing in with other recruits' footwear. We turned a clean, organized bay into a massive

heap of stuff. We even threw some things into the latrine. We took

canteens that were hanging on the bunks and emptied them out on sheets,

boots, mattresses, and pillows; as long as there was water in a canteen, it

became one with something on the floor. After me and the boys were

finished, we exited the bay, this time with 4th platoon standing at

attention directly outside their bay, solemn expressions all. They knew

they fucked up and they were paying for it.

There was only one Drill Sergeant that the Wolf Pack didn't like doing

CP (command post, the overnight watch on the company); it was Drill

Sergeant Z. She despised 3rd platoon and any chance she got she enjoyed

messing with us. We were all so close to graduation, and while other

companies' Drill Sergeants would take it easy on them for making it to

Blue Phase, we were still getting smoked, getting down in the dirt and

sweaty and still moving with a purpose.

One night, Drill Sergeant Zegler had CP, and called 3rd platoon down to

the formation area. It was around 2130, lights out was in half an hour, and we were the only platoon in the formation area since Drill Sergeant

Zegler wanted to "talk" to us.

She was joking with us and telling us she hated 3rd, and that her platoon

was better, and that since she was on CP that night, she didn't want any

trouble from any platoon. Drill Sergeant Zegler then asked us if we

wanted her to keep us company. Much of the platoon actually said yes,

and then she asked us if we wanted to play some video games. Again, the

majority ruled yes. Drill Sergeant Zegler made us run up to our platoon

bay, and when we all got inside, she told us to line up, one recruit in

front of the other. We did as she said, and with that, she told us to squat.

The Wolf Pack squatted in a long line, then Drill Sergeant Zegler told us

she had a controller, and we were the video game. She told us we were

playing a game like snake or centipede; when she told us "Walk forward;

left foot," the platoon moved their left feet forward as one unit. When

she said "Right foot," we all moved our right feet forward. Left, right,

left, right, repeat. We were doing this in a squatted position and we were

down in a squat for about 20 minutes, our legs burning and some recruits

stood up when she wasn't looking, just to ease the pain, even if for a second. We were a human centipede, in a manner of speaking. Left.

Right. Left. Right. Left. Right.

Drill Sergeant Zegler ordered us all to get up, and there was a very

audible sigh of relief from many of us.

"Alright Pri'ates, I'm going downstairs. Have a good night, and nobody

better *dare* knock on that door. You got that?"

"Yes Drill Sergeant!"

"Good. Lights out, Pri'ates."

And with that, the females went to their bay and the males turned out the

lights in ours.

There were a few days before graduation and the battalion was practicing

for the ceremony. When all the companies returned to their areas, Charlie

formed up in the company area and our Drill Sergeants told us to recite

the Soldier's Creed. After the booming creed was finished, one of the

Drill Sergeants from 1st platoon told his platoon to recite it. After they

finished, Drill Sergeant Price told 3rd to recite it. As we were doing so, he

gestured to us to get louder, and this went on until we ended up

screaming the Soldier's Creed. Drill Sergeant Jones of 1st commanded his platoon to about-face and shout the creed to us, 3rd platoon. Both platoons were eventually screaming the Soldier's Creed to each other, and the Drill Sergeants wanted their platoon to be louder, trying to "1-Up" the other. Then 2nd and 4th platoons joined in, their Drill Sergeants marching them right next to 1st and 3rd, and they screamed the creed as loud as they could. Drill Sergeant Price left-faced the last two ranks of 3rd so we could scream the creed at 4th and 1st at the same time. It was awesome, the Drill Sergeants were laughing and the platoons were trying to hold in their laughter to no avail. That was probably one of the best moments Charlie Company had throughout all of basic.

Graduation Day

Charlie Company of 1/13 Infantry Regiment finally made it to graduation. We did the ceremony for all our friends and family to see how far we've come, from civilian to soldier. After, our guests met with us in the company area for pictures and basic training stories, and we were even allowed to give our guests a tour of the bays. In a big way, I was going to miss this place; the Drill Sergeants, the camaraderie, the yelling, the intense smoking, the training...all of it. I will never forget the collective suffering during a smoking session or while out in the field, soaking wet, cold, and miserable. These moments are burned into my mind and are now memories, unforgettable experiences that I can tell everyone "I did it."

Drill Sergeant Price couldn't make it to the graduation ceremony that day because he had to leave for a mission, but before the ceremony, he shook the hands of each and every 3rd platoon member, telling us all good luck and best wishes in our Army careers.

Drill Sergeant Nelson won the Drill Sergeant of the Cycle title and award, and that felt great for all of Wolf Pack. One guy in our platoon earned an award for shooting Hawkeye on the range. He had the highest score possible and the highest in the company, followed by a few others from the remaining platoons. Also from 3rd, Wyche (the APG) and Daversa, (the PG), both won awards for maxing their APFT. We were Drill Sergeant Nelson's last cycle before she retired the Drill Sergeant campaign hat, and we were possibly her best cycle. At least that's what she says. It was a little emotional after graduation because the platoon pooled some of our money together to buy her an iPod dock as an end of cycle and final cycle gift, and also because she trained us hard but fair. We gave her superb soldiers with great potential and she gave us hard, exhaustive training---exactly what I hoped for. I wanted to get the best experience possible so I can be the best soldier I can, and with teachers and trainers that are Infantry, support MOS' with combat experience, and scary Drill Sergeants that know how to sweat a platoon, it was no Fort Benning, but I learned a great deal about properly wearing the uniform, marching a formation, and the correct fundamentals and body positioning of basic rifle marksmanship. I also learned how important teamwork is, and the benefit of having a strict attention to detail. To this day, I still remember everything my Drill Sergeants taught me, in addition to the new things I've learned over time. As it is said, learn all you can and apply what's useful.

Epilogue

After Basic Training, I went on to Advanced Individual Training, where I learned more computer skills and was trained on the various military radio systems. AIT was a step up from Basic in reference to maintaining and improving fitness, and learning to mix Army responsibilities and professionalism with civilian responsibilities and leisure. There was definitely a need to have a fine mingling of the two worlds which was important.

A small handful of recruits from Basic came over with me because they all had the same MOS. The company of the unit I was attached to was also Charlie, and again, this Charlie Company was strict and tough. Constant details, one of them consisting of cutting the grass---and if we didn't have the two weed-whackers or the one lawnmower, we had to either pull the grass out by hand or get scissors and cut it. Another fine detail (which I actually didn't mind doing) was raking the sand in the company area. It was very much like a Zen garden, with tiny pebbles strewn throughout, to the big rocks in the middle spelling out "C-ROCK." That is the name of the company, Charlie Rock, because we're a hard, tough company and you can't smoke a rock! So the Zen garden detail wasn't bad, although it was in the summer heat of Georgia, when the general army still had berets, before they integrated wearing the patrol cap. The damn beret kept your head cold in the winter and hot in the summer, which is why some soldiers despised wearing it. I will say though, that summer heat made it worse running or working outside. This one specific recruit had to do PT with my class in his tan shirt, ACU pants and boots in the hot heat and sun because he wanted to be stupid and watch porn in the middle of the bay in the middle of the day. Dumb-ass; he was sweating before we finished stretching.

I met some pretty cool people and made some good friends in AIT. I was going through this school in the New Army, where there were no more Drill Sergeants as the instructors or platoon sergeants, but whenever a buddy and I spotted a Drill Sergeant on base, we would immediately perk up and our Basic Training etiquette came back.

The platoon sergeants we had in Charlie were outstanding, and I learned much from them as well. My platoon won the top spot as the best platoon in the company (and it earned us a nice trophy to boot), and Charlie won second place behind Echo during field day (an enormous celebration with bbq, games, and competitions), which was ok but they took first place in Drill & Ceremonies, which was bullshit because we should have won that one, even though I will admit Echo Company looked real good in their movements and drill, and Charlie did have a couple moments where one Soldier mistook the next movement. Their judges of course were biased and voted 5/5 points for their company, whereas our company's First Sergeant would give his own company 3/5 points max, so that way no one can say he was biased. I took part in the D&C

96

competition; it was a bunch of fun. Seeing it on camera afterwards, it blew my mind how we executed a long list of facing movements and marching, including having all four ranks break formation, time their marching and direction perfectly, just to come back together in the same order we were originally. I underline that that D&C competition was an outstanding performance by all and was a great experience to be a part of. All the time training and practicing after school definitely paid off. During the week Charlie had Combatives training and we would go out to the dirt and roll with each other. We had a number of days where we held a free-for-all, a battle royale of sorts, thanks to our first sergeant. One on one, one on two, two on two, three on four, no matter the number, it was every man or woman for him or herself. That was always a bad-ass time.

I had a female buddy who was one of the guys; she was tall and beautiful and loved the high speed stuff of clearing rooms, combatives, shooting, all that. One of those days during battle royale in combatives, she and I rolled and she made me submit. Although I'm a guy, I'm humble and gave her a fist bump for her victory. Another time we rolled, I submitted her and that evened the score between us. Somehow, many people in Charlie found out about our matches and our tie, and one day we were outside after everyone finished rolling, and someone shouts, "Askedall and Miller!" That turned into everyone cheering and instigating, and one of the platoon sergeants says something.

"Oh yeah, I hear you two have a rivalry!" More cheering.

Miller and I step forward out of the mass formation of recruits and acknowledge we've been going head to head.

"Right here, right now, Miller and Askedall! Tie breaker! If Miller wins, all the males do fifty push-ups. If Askedall wins, all the females do fifty push-ups. Let's go!"

And with that, Miller and I square up as everyone forms a circle around us. I take her down, the crowd cheers, we roll, we get back up. We clinch, attempting to overtake the other. She takes me down and the crowd unleashes a loud "Ohhh!"

Now we're doing a ground game, I have my legs wrapped around her waist as she's trying to get a collar choke. She can't get it. She stands up and takes me with her, and I know what's coming, and so does everyone else watching. Remember, Miller's tall; she gets me at a good height and slams me onto the ground, dust flying and everyone going crazy. She picks me up and slams me again, the whooping cheers picking up again like a wave. After the third slam, I'm still attached and I buck my hips and push my legs out, pushing Miller to the ground. I detach, throw myself forward and tussle with her, the both of us trying to get the upper hand. I finally get her in a position where she has her back to me and I apply a blood choke. I squeeze tight…tighter…tighter…then Miller taps and I let go, everyone cheering. We both get up, hug it out, and absolutely everyone in our company gathering got down and started knocking out push-ups, regardless of who won.

There would be days when we would run with our First Sergeant too, and all of us took pride in doing so because he seldomly visited the company, being so caught up in his duties or us in our schooling. He always told us how he loved coming out and talking with us and conducting PT, because he had a passion for leading Soldiers and being there with us. He had excellent report with the majority of Charlie Company. I say the majority because not everyone liked his PT or even wanted to be there. Some just wanted to be discharged and be done with the military, not even out of AIT to become a real Soldier.

My dad flew out from New York to attend my Basic Training graduation and he was also there for my AIT graduation. Upon picking me up, we drove cross-country to California, stopping in different states along the way to visit family and friends. Integrating back into the civilian world and lifestyle was an easy transition, but I still kept the discipline and the prompt early morning wake-up. I noticed that I have acquired the strict attention to detail that has been drilled into me back in Basic; I notice every dried water spot on a bathroom or latrine mirror, every crumb on the kitchen counter and floor, and every speck of dust on a desk or countertop. I keep my room and my place of residence as clean and organized as I can, and since fitness in the U.S. Army Reserve is all on the Reservist, I do PT three to five times a week. Fitness is a way of life, and the more I apply it and learn about it and learn about proper nutrition, the more I like it and strive to stay fit and healthy.

My initial entry training was over, but my Army career was just beginning.

Poems by Paul Askedall

The Motivator

I met a motivator

One who pushes to succeed

By the time he made me push

I memorized the Soldier's Creed

This motivator is called a Drill Sergeant

He is not your friend

By the time he is through with you

You'll come out sharp in the end

Drill Sergeant, Drill Sergeant

My aim was good, you made it great

My mind was strong, you made it steel

I knew who I was, you put me in a better state

Bred for combat, you taught me what was real

Feel the Burn

Maxing the PT test, qualifying Expert are two standards you encourage

If you're not motivated he'll make you feel like dirt

"Moving Drill Sergeant!" now, not later

Need to get yourself in check?

Meet the Motivator

Return

Waiting to leave, I'm so cold

Nowhere has this chaos been foretold

Pedals falling to the hard ground

Brass is hot gathering all around

My heart beating is the only sound

Fellow Soldiers nowhere to be found

My eyes can barely see, the sun's aura blinds and tricks

I reload, running low on green tips

This is for my buddies in the combat zone

This is for my loved ones back home

Feel the Burn

Still breathing, alive, I'm fighting

Kill the enemy, kill the enemy; I'll paint the sand with their memories

Save me, save me, the silence puts me at ease

Violence I do not condone

Though I will fight, fight to hear my lover's soothing voice in my ear,

and be held in her embrace, alone

I will shoot, shoot to get away from this militarized zone

I will kill, kill to be the one to journey back home

Digging for Diamonds

Diamond push-ups tear your shoulders up, but

Please excuse me while I find some diamonds for this NCO,

as well as

Allow me to sound off loud and thunderous while I'm down here in the

front leaning rest

Doing my best

Push, down, up, halfway down and hold it

Feel the Burn

Loud, proud, no sympathy for the tired now

Still pushing, digging for diamonds

Not finding any, and still dirtying these hands

Earning my keep

Earning my sleep

Down, up, down, up

Toning this body as I sweat for equity

A smoking, they call it

Yeah, that's about right

Punishment for a wrong, I've learned what was right

www.ingramcontent.com/pod-product-compliance
Lightning Source LLC
LaVergne TN
LVHW011336080426
835513LV00006B/373